did you know?

A long time ago, cities set their time by the sun, calling 12 noon the time when the sun was directly overhead.

So even cities that were close together set their watches to totally different times.

Once the modern railway was built, and people were able to travel long distances in a single day, this way of keeping time led to lots of confusion.

Trains couldn't stay on schedule when clock time was so unpredictable. What was needed was a standard time.

So a very smart man named Sanford Flemming suggested we divide the Earth into 24 time zones, one for each hour of the day, so that trains, which were his life's work, could run on a schedule that everyone could share and understand.

Yay, Sanford Flemming!

Around the World
in the Blink of an Eye

Join us in our we(e)press community to meet our authors and illustrators, and chat with CODE staff and program people about their work in Africa.

Tell us what Africa program you're most interested in, and we'll allocate the proceeds from your book purchase there.

www.weepress.org

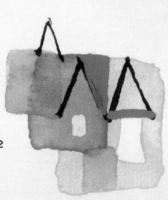

Copyright © 2010 Lou Hood

Illustration Copyrights held by their creators, under exclusive license to this work.

All rights reserved. No part of this publication may be reproduced or transmitted in any form or by any means, electronic or mechanical, including photocopying, recording, or by an information storage and retrieval system, without permission in writing from the publisher, except by a reviewer who wishes to quote brief passages in connection with a review written for inclusion in magazine, newspaper, electronic media or broadcast.

For further information: we(e)press
118 2nd Avenue
Ottawa, ON
Canada K1S 2H5
www.weepress.org

we(e)press is a Canadian Registered Charity # 80461 8353 RR0001

First edition April 2010
Proudly printed and bound in Canada by Friesens www.friesens.com

Design: Accurate Design (Louis Chan) www.accurate.ca

Photo credit: Dylan Leeder Photography, Alex Lukey Photography
Ben and Lou photo: Dianne Galus www.shootstories.ca
Ron Broda's Ottawa paper sculpture photographed by William Kuryluk

Library of Congress Cataloging-in-Publication Data
Hood, Lou, 1964-
Around the world in the blink of an eye : a truly global picture book / Lou Hood.

ISBN 978-0-9809728-1-8

1. Children—Social life and customs—Juvenile literature. 2. Manners and customs—Juvenile literature
3. Time—Systems and standards—Juvenile literature. I. Title.

GT85.H66 2010   j390   C2010-900492-2

This is a book that could only have been born in this modern and interconnected world of the Internet and email. That said, *Around the World* literally took us to some very remote and far-flung places, and in many instances we had to rely on helpers to assist us in both finding and communicating with local artists. Thanks especially to Joseph Johnston at ArteMaya galleries in San Francisco and Maria Stella Patera at Northern Images Arctic Co-Op in Inuvik, Mareligio Roque and Dave Millar. www.timeanddate.com was an invaluable resource throughout.

Thanks also to Diane Dufour, Angela Medaglia and Louis Chan of Accurate Design, Bank Street Framing, Friesens, Erik (skater boy) Kulakowsky CA, Mike Hood, Alan Morissette, Dianne Galus, Tim Barber, Bill Fox, Julie Danna, Christina Stefanski, Luke Swanek, the athletes of Bytown 171, and Scott Walter and Sean Maddox of CODE. Our appreciation to Lisa Lyons, Karen Boersma and Sheila Barry of Kids Can Press, and to educational consultant Kitty O'Flynn. Special thanks to Tom Jenkins for his encouragement.

Undying gratitude belongs to the 25 wonderful artists who so eagerly joined in this adventure. Theirs were some of the nicest "yeses" conceivable. This book owes its beauty and relevance to these great artists. They generously gave of their creativity and inspiration to illustrate a beautiful book for you, and to raise awareness and funds for education and literacy in Africa; they are all true stars.

Special thanks to my own true star.
This book has been so much a part of our life's story that it could only be dedicated to you, Charlotte.

Around the World
in the Blink of an Eye

a truly global picture book

By Lou Hood

cover by Chris Raschka

illustrated by 24 great artists from around the world

Have you ever thought about why we have daylight for part of the day, and night for the other part?

It's because this big blue marble we call Earth is constantly spinning!

Really... It's true!

And in 24 hours, or one full day, Earth spins one complete circle.

This makes it look like the sun is moving across our sky.

Hour by hour, how time flies! Let's go around the world in the blink of an eye!

So, part of each day every one of us will be on the sunny side of the planet for a while, and the rest of the day we'll be facing away from the sun.

And we created clock time to keep track of where we are in this daily rhythm of day turning into night turning into day again.

Here's the really cool thing...

When you are experiencing daylight, for someone else on the other side of the planet it's their night time!

Every hour of every day you are writing your life's story.

Can you find me on our trek around the world?

Ottawa, Canada

It's 7 o'clock in the morning
and the sun is rising in Ottawa.

It's time to wake up!

And at that very moment,
in the blink of an eye,
children who live somewhere else on
our great planet are eating their dinner,
or going to bed,
or riding the school bus,
or dreaming,
or just about anything you can think of
that happens to each of us each day.

*follow the hummingbird,
he'll show us the way
24 hours makes a day*

St. John's, Antigua & Barbuda

It's 8 o'clock.

The bus will soon be here to take us to school.

Buenos Aires, Argentina

Graciela and Marco like to share their snacks at recess. Marco likes alfajores the best, but the pigeons seem to prefer Graci's pochoclo!

São Paulo, Brazil

Mareligio and José love football. So does Marta.

São Miguel, Azores

Today our class went to Pico do Carvão. We saw an Açor and teacher told us our archipelago is called Açores after them!

Accra, Ghana

Naana would rather the quiet time on Mama's lap to playing football in the yard with her brother and sister.

Trieste, Italy

The breeze feels beautiful, doesn't it Oscar?

Oh, Oscar! All you ever think about is what's for lunch!

Alexandria, Egypt

"I can't waaaait until the big cousins get home from school, Giddo," Khaled told his grandfather. "Especially now that the little ones are done napping, and I don't have you and Tetta all to myself anymore."

14

Dubai, United Arab Emirates

Babba enjoyed the warm afternoon sun as he listened to Mohammed talk about his day.

"More tea?" asked Babba, when the boy paused long enough to notice the dhows sailing on sparkling Dubai Creek.

Lahore, Pakistan

Taj is proud to have such a fine new kite.

Almaty, Kazakhstan

She could hear Ermek's breath heavy beside her. He'd been in such a hurry, running along the ridge ahead of her, a bobbing bundle of kindling.

Ata was nearly home and soon they'd be around the dinner table.

Bayan's tummy grumbled with the thought of Schesche's special stew bubbling on the fire.

Vibol loves bubbles in his bath

Phnom Penh, Cambodia

Beijing, China

Lily secretly hoped that Father would win, so that she could play her game against him.

Seoul, Korea

I can still hear Umma's gentle voice, reading aloud to me. But she sounds far away and I know I must be drifting off to sleep.

The bumps in my mattress feel as though I am lying on the bumpy tiles of the rooftop.

I can hear Minsu's laugh. And I can feel the rumble of Shinjaakro's purr. I'm sure I could see the whole world from up here, if I tried to.

Yaroomba, Australia

That's the wonderful thing about words. They can paint a picture and just send it floating out into the night sky, carried on nothing more than the sound of a voice.

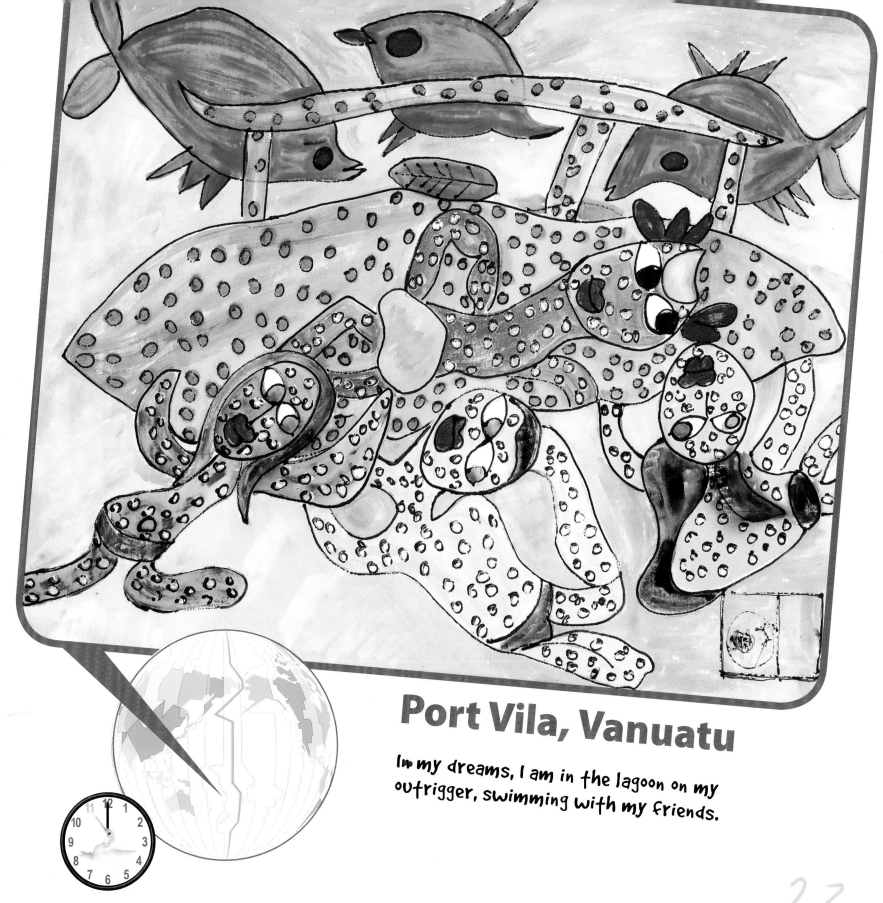

Port Vila, Vanuatu

In my dreams, I am in the lagoon on my outrigger, swimming with my friends.

Tarawa, Kiribati

Teweiariki sleeps under the stars.

Christchurch, New Zealand

George and Manu are dreaming about their favourite rugby football team.

They hope that they will have the strength and cunning of a taniwha in next Saturday's match.

Rarotonga, Cook Islands

Tomorrow I'll go to the market to help Auntie sell her fabrics. I wonder if Makea and her little sister will be at the fruit stand?

Anchorage, USA

Snowy branches shimmered as the Northern Lights danced across the sky.

Looking back at the cozy warmth of his bed, he laid the book beside his pillow, snuggled back under his blanket, and fell quickly back to sleep.

Inuvik, Canada

Genen helps on the trap line with Appa.

Comalapa, Guatemala

Like many children living in rural communities, Laura does not have the opportunity to go to school. Instead, she works in the house or in the fields to earn a little money.

Her story is all too common in this world. Sometimes the community is too poor. Sometimes the child is sent to work. Sometimes finding food is a greater need. And sometimes it is for no other reason than because she is a girl.

About our Artists

Ottawa, Canada
Ron Broda is a renowned children's book illustrator and paper sculptor. With over 15 years in advertising, art directing and commercial illustration, Ron combines techniques in paper sculpture and watercolour to create vivid illustrations with great detail and realism.

St John's, Antigua & Barbuda
Gilly Gobinet moved to the Caribbean in 1984. Her art, which depicts all aspects of life in the West Indies, often reveals her quirky sense of humour. Gilly's original paintings and books are highly sought after by local residents and visitors alike.
www.originalcaribbeanart.com

Buenos Aires, Argentina
Javier Gonzalez Burgos is one of Argentina's best-loved children's illustrators with more than 40 books to his credit. His digitally created whimsical characters are much beloved and have appeared in books and exhibits in numerous countries the world over.
www.illustrationartworks.blogspot.com

São Paulo, Brazil
Fernando Vilela obtained his Masters in Arts from University of Sao Paolo. Both writer and illustrator, he has received a number of prizes for his work, including a 2007 Bologna Ragazzi Honorable Mention. Fernando is a professor at the Tomie Ohtake Institute where he lectures on painting and book illustration.
www.artebr.com

São Miguel, Azores
Esperança Melo was born in the Azores, and has made her home in the village of Millbrook since 1990. She is a graduate of Sheridan College's Animation Program and a Graphic Design honours graduate from George Brown College. She has illustrated and designed several children's books and works in various media, including sculpting in papier-mâché.

Accra, Ghana
Meshack Asare is a prolific and multi-prize winning children's author and illustrator whose books have appeared in many languages and are celebrated throughout the world.

Trieste, Italy
Dilka Nassyrova was born in Kazakhstan and received her art education at Almaty State University. She has worked as a graphic designer and illustrator for magazines and in advertising and now lives and works in Italy. Dilka's work is filled with dreamy landscapes, fantastic furry beasts and odd doll-like, characters.

Alexandria, Egypt
Hatem Aly is a talented cartoonist, illustrator and new father, recently relocated from Egypt to Canada in 2007. A fine art graduate, Hatem has illustrated a number of children's magazines and picture books in numerous languages.
www.metahatem.blogspot.com

Moscow, Russia
Boris Kulikov graduated from The Institute of Theatre, Music and Cinema in St. Petersburg, Russia. Boris's books have been selected as "Best Books Of The Year" by The School Library Journal, Publishers Weekly, Child Magazine, and Time Magazine.
www.boriskulikov.com

Dubai, United Arab Emirates
Carla Hirst loves to find materials to fix onto paintings to give them a unique appeal and depth. A true hoarder, she keeps stuff everyone else throws away to enrich her paintings. Carla draws inspiration from exploring different cultures.
www.carlahirst.com

Lahore, Pakistan
Ambreen Butt was born in Lahore, Pakistan and completed her BFA at Lahore's National College of Arts. She moved to Boston to undertake her MFA at the Massachusetts College of Art, and resides in Boston to this day. She has shown her work widely and has been honoured with numerous awards and grants, including the Maud Morgan Prize from the Museum of Fine Arts in Boston.
www.ambreenbutt.com

Almaty, Kazakhstan
Assol Sas graduated from Almaty State University, Abai. She has illustrated six books, including a book by the Kazakh director and scriptwriter Yermek Tursunov about Mamluk soldiers. Her works are alive with romanticism and poetry, drawing inspiration from old Soviet illustrated books.

Phnom Penh, Cambodia
Stéphane Delaprée demonstrates a gentle insight, through his art, into everyday life in the homeland he adopted more than 10 years ago. Funny and colorful, his work is extremely popular and accessible. The universe he offers us in his work is happy and uncomplicated.
www.happypainting.net

Beijing, China
Ange Zhang studied theatre and design at Beijing's Central Academy of Drama. He went on to become a resident designer for the National Opera Theatre in Beijing. In 1989, he came to Canada, to the Theatre Department at the Banff Center for the Arts, and later the Stratford Festival. Ange was awarded the Bologna Ragazzi award in 2005 for his book *Red Land Yellow River*.
http://angezhang.googlepages.com/home

Seoul, Korea
Yangsook Choi grew up in Korea and moved to New York to study art. Selected as one of the most prominent new children's book artists by Publishers Weekly, Yangsook has written and illustrated many award-winning children's books.
www.yangsookchoi.com

Yaroomba, Australia
Tara Spicer is a young Australian artist full of passion and energy. Strong colors, romantic ideals and the odd splash of humour are Tara's own unique signature style.
www.nirvanavisions.org

Port Vila, Vanuatu
Aloi Pilioko is one of the South Pacific's best-known contemporary artists and a true ambassador of Oceanic art. Along with Nicolai Michoutouchkine, the two artists have created a public gallery and museum. They recently published a book celebrating their 50 years of artistic creations in Oceania.
www.nicaloi-aloi.com

Tarawa, Kiribati
Cait Wait graduated as a secondary school art teacher in Adelaide, Australia. Seeking to discover a sense of belonging, she began her travels around Australia and to the central pacific islands of Kiribati, where she worked creating hand painted fabrics, silks and garments, and taught art. She currently resides in Darwin.
www.caitwait.com

Christchurch, New Zealand
Gavin Bishop is a renowned children's book author and illustrator who has published some 30 books that have been translated into eight languages. The Storylines Gavin Bishop Award aims to encourage and publish emerging talent.
www.gavinbishop.com

Rarotonga, Cook Islands
Kay George moved from New Zealand to the pacific island of Rarotonga, Cook Islands in 1988. Together, Kay and her husband Cook Islands artist Ian George run their *The Art Studio* gallery. Kay holds a Master of Art and Design through AUT University, Auckland.
www.theartstudiocookislands.com

Anchorage, USA
Amy Meissner is an accomplished picture book illustrator. A watercolourist, Amy graciously offered to return to her roots in textile design to produce her piece for this book. Amy recently completed an MFA in Creative Writing and has undergraduate degrees in both Art and Textiles.
www.amymeissner.com

Seattle, USA
Kevan Atteberry's best known work may well be the most annoying cartoon character in history… the Microsoft paperclip helper, Clippy! Kevan has lent his whimsical style to a host of clients, including Amazon.com, Microsoft and Oracle.
www.oddisgood.com

Inuvik, Canada
Mary Okheena started drawing and printmaking at the Holman Eskimo Co-operative as a teenager. She is part of the third generation of organized graphic artists in the Canadian Arctic. Mary has illustrated a book of Arctic Folktales entitled *The Dancing Fox*.
www.northernimages.ca

Comalapa, Guatemala
Paula Nicho Cumes is unquestionably the most important living Maya woman artist in Guatemala today. Paula, a Tz'utuhil Maya, began as a weaver, only later learning to paint. Her themes are among the most original of any of the Maya artists, often painting themes that come to her through her dreams.
www.artemaya.com

Learn More www.weepress.org/aroundtheworld

read. learn. live.

At we(e)press, we love the idea of giving children the opportunity to help other children. You can be proud to know that the money raised from your purchase of this book will help children in Africa in their effort to build a better life for themselves through literacy. The sale of this book will raise funds in perpetuity for CODE's work with local African organizations that empower children to learn. For fifty years CODE has been delivering an essential and enduring sustainable development solution. Your purchase of this book will support libraries and teacher training, as well as national and local book publishing in about 20 languages in Africa.

Learn more about CODE at www.codecan.org

Help direct your funds to the we(e)press + CODE program of your choice www.weepress.org/uchoose

friends of we(e)press

This book was made possible through the generosity of:

   bluesky strategy group inc.

The Bitove Foundation